iOS Programming Guide

What Every Programmer Needs to Know About iOS Programming

By: Jason Scotts

TABLE OF CONTENTS

Publishers Notes

Disclaimer

This publication is intended to provide helpful and informative material. It is not intended to diagnose, treat, cure, or prevent any health problem or condition, nor is intended to replace the advice of a physician. No action should be taken solely on the contents of this book. Always consult your physician or qualified health-care professional on any matters regarding your health and before adopting any suggestions in this book or drawing inferences from it.

The author and publisher specifically disclaim all responsibility for any liability, loss or risk, personal or otherwise, which is incurred as a consequence, directly or indirectly, from the use or application of any contents of this book.

Any and all product names referenced within this book are the trademarks of their respective owners. None of these owners have sponsored, authorized, endorsed, or approved this book.

Always read all information provided by the manufacturers' product labels before using their products. The author and publisher are not responsible for claims made by manufacturers.

© 2013

Manufactured in the United States of America

DEDICATION

This book is dedicated to those who work every day to find new ways to do things in the technological arena.

CHAPTER 1- WHAT IS iOS SDK?

The iOS SDK is a software development kit (SDK) created by Apple to develop and test applications intended for the iOS operating system on mobile devices. Testing occurs with a simulator known simply as iOS Simulator. The iOS SDK was formerly known as iPhone SDK before development of the iPad and the iPod Touch, which also run using iOS. It was first announced in October 2007 by Steve Jobs and made available to third-party developers in March 2008. The SDK itself is free, but installing an app onto a supported iOS device requires users to be a paid member of the iOS Developer program, which costs $99 per year.

The iOS SDK has been included in the XCode suite since the release of XCode 3.1. The use of iOS SDK requires the installation of XCode. XCode is an integrated development environment (IDE) for use with Mac operating systems. It is a comprehensive collection of programs, tools and instruments used for development geared toward both Apple operating systems. It is used to code, build, debug, run, install and test applications. It also comes with iOS Simulator.

Both iOS and OSX applications are written in Objective-C, C++ or C languages. iOS uses Core Location as its software framework amid the programming framework, Cocoa Touch. Core Location allows for applications to retrieve the device's heading from a built-in magnetometer or digital compass and can also acquire raw heading data. This unique framework has been part of the iOS SDK since its initial release. Cocoa Touch is based on Cocoa with emphasis on touch-based gestures.

New versions of iOS SDK are typically released with each new iOS version, both major updates and minor updates. Beta SDKs are typically released to developers before their actual release to the public. Beta testers assess compatibility between current versions and new features. The current iOS SDK is version 6.0 and it is found in XCode 4.5 and higher. It requires the use of an Intel-based Mac system with OSX 10.7.4 Lion or later. Microsoft Windows and older versions of OSX are not yet supported for the use of the SDK.

Features of iOS SDK

iOS uses variants of the XNU kernel that is also used in OSX. BSD sockets for networking are also a shared feature of both operating systems. The tool chain for developing for iOS is based on XCode, making it sensible to include the iOS SDK in the extensive XCode suite. The contents specific to the iOS SDK include Cocoa Touch, which includes multitouch controls and events, Accelerometer support and camera support, along with i18n localization and hierarchy view. It also includes OpenAL, OpenGL ES, audio mixing and recording, image file formats, Quartz, video playback and core animation in its media set.

The iOS SDK core services set includes networking, an embedded SQLite database, threads, Core Location and CoreMotion. The Mac OSX kernel set accommodates TCP/IP, power management, file system, sockets and security features. iPhone Simulator is a program included in order to test an application using both the look and feel of an iPhone, iPad or iPod Touch on the developer's desktop without needing to install the program on the iOS device. It was originally called the Aspen Simulator but was renamed with SDK Beta 2.0. It is not an emulator because it actually runs code for x86 target not an ARM.

Use of the iOS SDK

There are well over 500,000 apps available in the iOS App Store and the vast majority were created by third-party developers using the iOS SDK. Developers are allowed to set any price that they want for their app and they receive a 70% share of the sales. A large amount of Apple's revenue is made in this way.

There are some restrictions imposed on application development using the iOS SDK, which are outlined in the Terms associated with the membership in the paid Mac Developer program. An example of this is running Java, which is currently outside of the bounds of the agreement. .NET frameworks cannot be installed and iOS does not support Adobe Flash.

Current iOS SDK

In September 2013, XCode 5 was released along with iOS 7. XCode 5 came equipped with the iOS 7 SDK. The LLVM compiler is capable of building 64-bit apps for the new iOS. It also came with a new Automatic Configuration allowing iCloud and Game Center configuration with one click. The Preview Assistant is capable of displaying in a UI in both portrait and landscape view for iOS 7 as well as the older iOS 6. New Debug Gauges featured in XCode show developers' important factors such as CPU usage, iCloud usage and more in one convenient location.

The iOS 7 SDK also comes with a number of multitasking enhancements. There are three new APIs for developers to take advantage of: Background Fetch, Remote Notifications and

Background Transfer Service. The new APIs allow better utilization of hardware and an improved user experience.

Background Fetch allows app designers to create applications that can passively download content while remaining in the background. This is good for news apps that get their content from the web, avoiding lengthy update times if the app had not been opened in a while. Updates can be programmed to take place at regular intervals. Remote Notifications is similar to the Background Fetch API. It utilizes Push Notifications to tell users when new content is available without downloading the content or constantly checking for new content.

The Background Transfer Service API gives applications the ability to have unlimited transfer time whether they are running in the foreground or in the background. Background Transfer Service can also work alongside either the Background Fetch API or the Remote Notifications API as needed. The system becomes responsible for managing data transfer and wakes up applications when necessary. This achieves greater energy efficiency through discretionary or non-discretionary transfer options. With these three new APIs in the iOS SDK developers now have more freedom to optimize their apps and make them even more user-friendly.

CHAPTER 2- HOW DO YOU GET STARTED WITH IOS- C PROGRAMMING?

As we go about our day-to-day duties of using our phone and the various bells and whistles that they come with, we don't usually think about a couple of things. Have you ever thought about how apps come to be on your I-Phone or I-Pad? Anything ever make you wonder how they just 'mysteriously' appear to your phone and electronics?

Did you know that involves images, touches, and gestures? There are different steps for an app to become and app on your device. In this article we will attempt to give a brief overview of the steps in programming of IOS, and in the end, give you all, the reader, an idea of what needs to happen in case you decide to participate in the journey of programming IOS.

Let's look at images and how they come to be in the world of IOS; in an apple product, there is something called a core image processing and analysis technology. In order to understand what Core imaging is, one would have to know about programming and all that interfaces with this application. Much like a computer that most everyone else uses, it is basically what runs behind the scene of the operator or the user. What is under the hood of a car or computer if you will.

From a high level perspective, core imaging provides the following; Access to build-in image process filters, the ability or capability for detection. It provides automatic image enhancements and the ability to chain multiple filters together to create unique and customary effects.

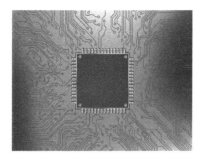

If you go on the website for Apple Inc. you will learn that there are more than a dozen categories of filters. Some are designed with the intent to attain artistic results. For example, style and halftone filter categories while others are more for optimal for fixing various flaws, flaws like color adjustments and sharpen filters.

Core imaging can detect the quality of images and determine what set of filters are needed to correct the image such as hue, contrast, red eye. Core image can also detect the human face features in still images as well as track them over in video images. The ability to know where the faces are in the image will help determine where to place the filter.

According to Apple, core image comes already with dozens of built-in filters to support the processing in your applications.

Touch technology can be found in the Cocoa Touch Frame work of Apple. Within the frame work of Apple there are many patterns; however they were built with a special focus on touch based interface and optimization.

This touch based technology is based in the "Objective C" of the Apple programming. This technology has the amazing ability to operate at very fast speeds. For this you would have to know the basic understanding of a program called C ++. Because of the compatibility of object C it will be easy to mix C and even C++ in the Touch applications.

Gestures or often known as Xcode gestures and Keyboard Shortcuts, can be used to simplify and or enhance your experience in using the Apple products.

There are Multi-Touch gestures and three of them are used and applicable within Xcode.

1. Three fingered swipe; up and down in the source editor. This switches from the source file and the associated header.
2. Two fingered tap; the main user interface element or area to open a contextual menu that element or area, or on an object in interface Builder to open an objects connection dialog.
3. Two fingered swipe; up/down or left/right, to scroll vertically or horizontally respectively.
4. If you go to the hyperlink https://developer.apple.com/library/mac/docume ntation/IDEs/Conceptual/xcode_help-command_shortcuts/MenuCommands/MenuCom mands014.html

You will find a table the gives you all of the Menu Command Shortcuts (By Menu).

Moreover, when developing an app there are certain gesture recognizers that can be added in order to simplify particular actions. The feature is supported by UIKit framework, a predefined gesture provider which works in favor of users reaching their high expectations. Besides predefined gestures you can create personal actions of your own by enlarging the existing list.

Programming a gesture implies attaching that certain movement to a view or a target. However, a view can have more than one gesture attached to its event handling. Due to this fact each gesture has its own personal action:

1. Tapping
2. Pinching in and out
3. Dragging
4. Swiping
5. Rotating
6. Long press

These are six predefined gestures, tapping goes for select, pinching in and out stands for zoom and so on. On the other hand, programming a gesture involves discrete or continuous actions either sending single alerts to its target or permanent messages if gesture is taking longer.

Creating new gestures may fail user expectations, in conclusion it is advised to use predefined gesture recognizer that meet user's perspectives. Most of the apps implicate minimum size buttons which cannot understand certain commands so it's safer to program your app with existing predefined gestures.

There are three steps to be followed when adding a gesture recognizer to an app:

A first step includes creating a gesture recognizer sample adding a target, an action or other specifications.

Assign an action to a particular view.

Implementation of gesture recognizer within an app can be made using Xcode Interface Builder or by pro-grammatically adding it to the app. If using Xcode Interface Builder simply drag the gesture from the library and attach it to a target view. If creating a new one it will be necessary connecting the gesture to an action method. When programming a gesture recognizer for a specific action within the application it's necessary to specify the the target object and desired action.

Responding to gesture recognizers refers to two different types:

Discrete gestures – represent recognizers connected to an action method. Each one of the gestures has different properties helping the whole action develop in a certain way, depending on targeted view;

Continuous gestures – represent recognizers which permit apps to respond gestures simultaneously.

CHAPTER 3- WHAT IS XCODE?

XCode is a development environment created by Apple for OSX and iOS. Specifically, it is a integrated development environment or interactive development environment (IDE), which means it contains a suite of software including a source code editor, build automation tools and a debugging program. The first release of XCode was in 2003 and the current stable release is version 5.0, which is free from the Mac App Store for OSX Lion and OSX Mountain Lion. Previous versions as well as previews of upcoming releases are available for registered developers.

General Features of XCode

Capable of building universal binaries, XCode works with the Mach-O executable format. Mach-O is short for Mach object, which is a file format for executables, object code, dynamically-loaded code, shared libraries and core dumps. Mach-O allows for fat binaries, in turn allowing for software to run on both Intel-based and PowerPC platforms. XCode also uses iOS SDK to both compile and debug applications that run the ARM processor. Also for compiling it includes the GNU Compiler Collection and the llvm-gcc compiler. Back-end debugging is accomplished with LLDB Debugger.

Apple's WebObjects tools and frameworks were also previously included in XCode, making it capable of building Java web applications. As of version 3.0 WebObjects was replaced by WOLips. XCode includes Instruments, a GUI tool. It runs on top of DTrace. Also included is Interface Builder, which can be used to create graphical user interfaces.

Supported source codes are C, C++, Objective-C, Objective-C++, AppleScript, Python, Java and Ruby. Third parties have also added support for GNU Pascal, Free Pascal, C#, Perl, Ada and D. It supports programming models for programming models Cocoa, Carbon, and more.

XCode Version History

XCode 1.0 was based largely on Project Builder. 1.5 improved on the code completion and debugger. Then with the release of Mac OSX 10.4 Tiger came XCode 2.0, which included the Quartz Composer, Ant support and the Apple Reference Library tool. XCode 2.1 was the first version to support universal binaries as well as Shared Precompiled Headers, conditional breakpoints and watchpoints, along with unit testing targets. 2.5 was the final version released for Tiger.

Mac OSX 10.5 Leopard saw the release of XCode 3.0, which included a number of notable changes from the 2.x series. It included an early version of Instruments, and then called DTrace, refactoring support, Objective-C 2.0 with garbage collection and context-sensitive documentation. It also supported Project Snapshots and Message Bubbles to show build error debug values alongside the code. XCode 3.1 was significant because it was the first version to be included in the iPhone SDK and it could target iOS 2.0.

XCode 3.2 was released with OSX 10.6 Snow Leopard and was incompatible with earlier versions of OSX and also dropped support for targeting versions earlier than iPhone OS 3.0., but could still target earlier versions. The last version available for Snow Leopard was 3.2.6 and it required free registration with Apple's developer website.

At the Developer Tools State of the Union address in June 2010, Apple announced XCode version 4.0. It consolidated all of the editing tools and Interface Builder. The software was free only to users that were paid members of the Mac Developer program and iOS Developer program. Both programs were a $99 per year membership. In the Mac App Store it was available for $4.99. Version 4.0 dropped the support for older systems. It also dropped its support for PowerPC and SDKs for Mac OSX 10.4 and 10.5 and all iOS SDKs older than 4.3. It could still target all of the older platforms.

4.1 was made free for all users with the release of OSX 10.7 Lion. In August of 2011 XCode 4.1 was also made available on Snow Leopard for paid Developer program members. It was the last version including GCC instead of LLVM GCC. XCode 4.2 included a number of other improvements including storyboarding and was the last version supporting Snow Leopard, but again was only available to paid members.

XCode 4.3 was released in early 2012 as an application bundle from the App Store. The menu was reorganized. Version 4.3.1 came out soon after to accommodate OSX 7.1. 4.3.2 was released soon after that with an improved iOS Simulator. Then 4.3.3 updated SDK for Lion. The next version, 4.4 was released in July and could run on both Lion and Mountain Lion. It also supposed automatic synthesizing, new Objective-C features and more. XCode 4.4.1 included bug fixes for 4.4.

Along with iOS 6, XCode 4.5 was released with support for iOS 6 and the 4-inch Retina display associated with the iPhone 5 as well as fifth generation iPod Touch. It also brought additional Objective-C features and other iOS support. 4.5.1 improved stability and and fixed a number of bugs. Soon after 4.5.2 was released to support the iPad with retina display and the iPad Mini. Then 4.6 was released with iOS 6.1.

XCode 5.x Series

The current XCode 5.x series expands upon the 4.x series and also includes support for OSX 10.9 Mavericks. It supports iOS arms armv7, armv7s and arm64. It has been integrated with Cocoa and

Cocoa Touch to create an extremely intuitive interface. Auto Layout adjusts to screen size and orientation and includes simple icons. It can even fix layout issues for you. This allows for easy transitioning between composing, designing and debugging. Live Issues finds bugs and reveals them as code is being typed using the LLDB debugging engine. There are also new debugging gauges to monitor CPU activity and more. The new Test Navigator allows for extensive test-driven development.

XCode 5.0 contains new bots for automation to adapt for continuous integration with the OSX Server as well as Mavericks. The bots can continuously build apps, execute and test them. There are also new menu options such as Source Control and new Accounts preferences. It is easy to monitor repositories as well as host repositories for a development team.

The IDE of XCode 5.0 includes Assistant Editor, Asset Catalogue, Open Quickly using Cmd-Shift-O, Source Editor, Interface Builder, OpenGL Frame Capture, Snapshots, Refactoring, iOS Simulator, Integrated Build System, Complete Documentation and a number of instruments as well as addition tools to create a comprehensive platform and environment.

CHAPTER 4- HOW TO DESIGN INTERFACES IN iOS

The design process should take up most of your time. That time will be used to familiarize you with the iPhone. Learning about it and using it will usually trigger an idea of a design that you would like to create in your own application. Once you have conceptualized the design, it's time to make a prototype and plan everything out with pen and paper. Then finally you're ready to take an average application and make it spectacular.

Becoming Familiar with iOS

The iPhone has a human interface guideline that everyone should read thoroughly before designing an interface with IOS. It will discuss interactions, context and constraints. Also it's best to study every apple application that ships with the phone to get a feel for how they design in interface.

Before designing in interface, think of ways to make your application appeal to the targeted users. Majority of users view their phone a few feet away at a hip level. They want access to the latest information and for everything to be updated. Your app should be intended to fix a problem and fulfill a need to customers.

The design has to work with the primary input methods. There isn't a mouse so now hit targets must be a minimum of 22 pixels vertically or 44 on average square. Embracing the new interaction methods will keep everything user friendly. The second input method is the keyboard. Since the keyboard isn't full sized, minimizing the amount of text entries will help users. Assuming what users would have to enter and providing it as a preset text would eliminate frustration. Having multiple options available so that users don't have to type it on their own will give your design an advantage over other apps on the market.

Conceptualizing iOS

Discover the purpose of the app. Once you have the purpose, filtering the features that aren't of any use to your design is key to having a successful interface. Select the bare minimum of features. Pick the few key features that are most frequently used by the intended users that are appropriate for mobile context. The three types of application types are productivity, utility and immersive.

To design the best interface you must understand the basics. In the iPhone the experience is a linear hierarchy experience. For native applications the title area should showcase what screen the user is currently on. Back buttons go on the top left and the add button goes on the top right. There are exceptions to the rules but the majority of users find this layout the most natural.

The primary way to view information is by creating a list. There are plain tables and group tables. Plain table views are recommended for items of more than 20 on a list. They give unconstrained data sets, categorization and quick navigation. Grouped table views give users more visual stimulation. They are easy to customize also. They deliver constrained data sets, grouping of like items and are graphically richer.

Tool bars are the containers for the verbs or actions inside of the application. It should be specific to the screen above it with familiar icons without textural labels because users will confuse it for a tab bar. Tab bars serve as a high level of abstraction inside the application. Tab bars are normally permanent and give users the option to customize the application. It increases efficiency when used to switch between modes, categories, collections, objects and instances. It's not a toolbar alternative. Order the tabs by the frequency of use. Allow users to launch your application and receive new, fresh data that might interest them.

The Aesthetics

The aesthetics of the application shouldn't overload the user. Avoid cramming too much on the screen. Make sure the text has enough padding, both horizontally and vertically. Data should be prioritized with the most frequently used items at the top and the least used going towards the bottom. Have data that offers critical functionality and gets frequent use.

Group related items with group tables. Use UI metaphors that are standard, built-in controls because the completed set is available, it's familiar to users, visually consistent and optimized. It's important to make the buttons usable on the iPhone so round the corners.

Using Pencil and Paper

Design your interface on paper. Create preliminary sketches of potential UI's for the application. Prototyping on paper will show if the app is user friendly and follows the guidelines. When creating the prototype, configure to the new UI paradigms of the iPhone.

Think about the accidental touch user input. Consider the outside lighting conditions when designing.

Finalize It

Once the standard UI is clean and ready to finish, think about if photos and imagery can enhance the experience. Adding animations for actions add excitement to the interface. It enhances a task or concept and it can also clarify a process or concept.

The goal is to stand apart from competitors. Don't leave your interface in its clean and standard condition as this allows competitors to copy your design. Think of little ways to make your UI unique to only your brand, that way any coping can be easily detected and dealt with.

iOS is constantly improving and making it easier to create interfaces. So depending on when your app is created it can become even easier than before. When designing the interface in iOS always craft an application definition statement. Everyone who plays a part in designing the interface needs to abide and honor the statement so that no one tries to add to it. Iterate on paper often because it will save development time later. Repeat this cycle until it's ready to be released. Releasing a buggy or dysfunctional app will be deleted and users don't think to download an app twice so get it right the first time. Make sure it's something great upon the first release.

CHAPTER 5- HOW TO DEVELOP ANIMATIONS AND VIEWS IN iOS

When asking information technology (IT) graduate school students to explain how they develop animations and views in iOS, the response was "do you want it in Geek or English?" In fact, the thing that usually does a non-Geek's head in is trying to speak the same language as an IT professional. The result is confusion because most computer users do not use popular iOS terms such as SDK, jail breaking, unlocking the SIM lock or the iOS kernel. Simply put, the best way for a layman to understand animation development and views in iOS is to know that it is all about creating computer interaction with touch, swipe, tap and multi-touch interface by rotating an image or art in some way to create movement.

In general, the view from IT experts is the creation of animations is somewhat complex for non-tech types, but is also straightforward because it does not mean drawing or writing code. Animation techniques are not linked to traditional art methods such as drawing or painting. In fact, the use of core animation is all about changes a computer app designer uses to "animate changes" that a user sees when activating an app. Thus, this type of animation has nothing to do with the classic Walt Disney method of drawing images that move on screen.

Creating Little Images That Move

For instance, the interesting little picture images - on an iPhone or today's ever popular applications or apps - feature millions of iOS applications that were once the brainchild of a tech designer wanting to animated the Smartphone search experience with a piece of art. In fact, the typical musical note clip art that has been around for decades is now an iOS application for the iTunes store, or a clip art image of a telephone is now, guess what, an iOS app for a cell phone call. While most IT experts enjoy demystifying how they write apps for iOS – when using either a 32-bit or 64-bit ARM architecture – they freely admit that the techno-speak may be over the heads of non-Geeks who simply want to understand what the fuss about iOS is all about.

According to a college textbook chapter that explains methods for developing various "animations" in iOS, one needs a strong background in general computer technology. The textbook states it is a fools game to try and decode the same tech speak that the late Steve Jobs used with his Apple hardware and software designers. The solution is to work with an IT professional who can take your app ideas, and create the illusion of animation by using what tech experts call interface control elements that include buttons, switches and sliders that app users to get those messages, photos, maps, notes and other data animated on their smart devices and screens.

Breaking the Code of iOS

First you have to know what the difference between tech-talk and everyday language that people use to explain things. When asked how to develop animations and views in iOS, an IT professional will tell you that animations is simply a "tool" that tech designers use to give various "screens" the illusion of some cyber realm or experience. The same IT expert commented online that the layman might get confused with the iOS as the once touted iPhone OS. To end the confusion, the IT expert said iOS is simply a short tech-speak term for mobile operating system.

History of iOS Reveals Animation Methods

Because Apple states that its animated apps have been downloaded more than 70 billion times, there is huge interest in producing more animated next generation iOS screens that attract mobile users who want to be entertained when searching information. The use of animation is linked to how human beings naturally behave online with apps designed for reading data and browsing. Also, it is important to note that Apple has licensed the famed IOS trademark. Thus, anyone attempting to create like animated apps using Apple's IOS animation methods may be opening themselves to a law suit from one of the most powerful companies in the world.

How A Few Lines of Code Creates Animations

The act of changing how a user views an application - with movement or transitions to replace one image with another – is a process allows humans eyes to view images that are much more than just smoke and mirrors. In fact, a few lines of computer code can make a smile face smile at a user, or a map app open to a country scene and roadways where a user wants to go. Still, most core animation simply provides "support for animations," explained an IT expert with a big grin because this techno-speak often gets very confusing and even funny.

In general, the techno-speak for creating animations in the digital realm means animating something that a user views. The changes that a user views on an app, for example, is all about modifying some image seen on a screen so a user can view the image or app from a vantage point or smartphone that simply changes the size of a system. So, animation simply means the ability to rotate, translate or transform images into everything from 3D to a simple background color.

The process of animated view transitions include:

- How to animate size
- How to make something on screen transition to another screen or image
- How to control the behavior of data being displayed

In general, IT experts use something called an animation blocks as "code" for the use of various code that makes animated changes that users see when they touch an app and it dances, makes sounds or changes colors. Also, the used of block-based methods are taught at many community colleges, colleges and universities under select computer course work that focuses on these methods.

While there are many tech-savvy people out there in cyberspace who can understand animation blocks, the process still requires proper training.

CHAPTER 6- IOS PROGRAMMING- HOW TO CREATE IMAGES, TOUCHES AND GESTURES

Images

First start an Xcode project. Once you have an Xcode project set it on a single view application, have storyboards and automatic reference counting and device set to iPhone. Then go over to the viewcontroller.h file. Now on the line @interfaceviewcontroller: add the code UIImagepickercontrollerDelegate, UINavigation (controllerDelegate>

You want to create two UIImagepickercontrollers and name both differently.

UIImagepickercontroller .picker;

UIImagepickercontroller .picker2;

Also add UIImage .image;

IBoutlet UIImageview .imageview:

The way to make it all work is with two buttons, one to make the camera and one to use an existing photo from the photo library!

- (IBAction) Take Photo:

- (IBAction) Choose Existing:

Now move over the viewcontroller.m file and work on implementation add

```
- (IBAction) Take Photo{
```

picker= [[UIImagePickerController alloc] init]: picker.delegate=self

[pickersetSourceType: UIImagePickerContollerSourceTypeCamera];

[self presentViewController:picker animated: YES completion: NULL];

[picker release];

}

Next you'll need to add the actions

```
- (IBAction) Choose Existing{
```

Copy and paste previous actions but add a 2 after picker. So everything that was picker is now picker2. Also after you copy and paste where it says ControllerSourceTypeCamera change Camera to Library. It should now read ControllerSourceTypeLibrary.

Next it's time to add more codes

```
-           (void)imagePickerController:(UIImagePickerController.)
pickerdidFinishPickingMedLawwithInfo: NSDictionary.)info{
```

image=[info objectiveForKey:UIImagePickerControlleroriginalImage;

[imageviewsetImage:image]

[self dismissViewControllerAnimated: YES completion: NULL];}

```
-(void)                          imagePickerControllerDidCancel:
(UIImagePickerController.)picker{
```

[self dismissViewControllerAnimated: YES completion: NULL];

Those are all the codes that need to be manually added.

Then click on ViewController.xib file and drag some buttons. Drag two round rectangular buttons to the top of the screen. Double click on them to rename them take photo and choose existing. Then select a button click on first responder, then connections inspector and marry the function to the button while adding the "touch up inside" function. Do this for both buttons. Drag and drop the imageview to the screen as well. Adjust the size of the imageview to better see the images once it has pictures to display. Next control click "file's owner" and drag and select imageview. Now you'll want to run the application. The buttons should work on your mobile device. That's how to program images in iOS with Xcode.

Now that you have images you can add filters to them with Core Image.

Pinch Gesture

Start a new project in Xcode using a single view application with automatic reference counting. Don't check "use storyboards or unit tests". First thing you're going to do is build your interface by clicking on ViewController.xib. Next add an Image View and scale down the image view so that later on you can zoom in and out. On the right hand side of the screen, check off "User Interaction Enabled" and "Multiple Touch". Now you need an image, so drag an image to "supporting files". Then make sure to check off "copy items into destination group's folder", and then click on "finish".

To make the image appear in the image view on the right side of the screen next to "Image" open the drop box to retrieve the image. Click on the "pinch gesture recognizer". Then drag it on top of the image. Make sure it was added by checking your objects list. You can click on pinch gesture in the objects list to double check that it's connected to the photo/image.

Next you need to create a method that can handle the pinch gesture, so open up the editor and right click and drag the pinch gesture recognizer into the header file. Next create an action and name it.

Staying in editor it should read

@interface ViewController: UIViewController

- (IBAction) "the name you named the action": (id) sender;

You'll want to make (id) sender more specific, so change that to (UIGestureRecognizer*)

Copy codes and add them to the implementation file. Replace the original code with the new more specific code. Returning back to the header file, once again right click and drag from image view to the header file. This time making the connection an outlet and naming it something different.

The zooming in and zooming out code to add to the "ViewController.h" is listed as follows:

```
CGFloat lastScaleFactor=1;

CGFloat factor= [(UIPinchGestureRecognizer*) sender scale];

if (factor >1) {//zooming in

imageview. transform= CGAffineTransformMakeScale(

lastScaleFactor + (factor-1),

lastScaleFactor + (factor-1));

}else{//zooming out

imageview.transform= CGAffineTransformMakeScale(

lastScaleFactor *factor;

lastScaleFactor *factor);

}

if(sender.state==UIGestureRecognizerStateEnded)

{

if(factor >1)

lastScaleFactor+=(factor-1);

else

lastScaleFactor *=factor;
```

Now you're able to gesture pinching and zoom in/out an image. To add other gestures such as the tap gesture, swipe gesture and the pan gesture the process is similar. You always start a new project in

Xcode and you need an image view. The only changes will be in the action codes and the outlet codes. But once you practice it will become easier to master.

Touches

To add buttons to touch and create actions onto the screen, first start a new Xcode project and name it. Go to the header file and add code:

@property (nonatomic, retain) IBOutlet UILabel*ourLabel;

- (IBAction) buttontouch: (id) sender;

Then go to implementation file and add @synthesize ourLabel=_ourLabel; towards the top of the screen under @implementation. At the bottom of the screen before @end add code

- (void) buttomtouch: (id) sender {

[self.ourLabel setText: @ "Hello Button! :) "];

Now when you got to open ViewController.xib you'll add a label by dragging it onto the image view. Adjust the size of it as you see fit. Add another label of rounded corners and title it "our button".

Click on "files owner" and where it says ourLabel marry it to the label. Then click on buttonTouch and marry it to the ourLabel button and open the drop box to select "touch up inside".

Then lastly run the application to make sure it works. When it says "Build Succeeded" that means it works and the iPhone simulator should appear. Now you can touch the screen on the button you added and it will work properly. When you click on "our button", "Hello button! :)" will appear at the top of the screen.

These are some of the easiest and most basic ways to add touch, gestures and images in iOS.

CHAPTER 7- HOW TO BUILD TEXT IN iOS PROGRAMMING

First you have to know what the difference between building text in iOS programming with something like Xcode 4; while using your computer and an online website tutorial. In turn, you can do the same thing with expert instruction offered by information technology (IT) experts who teach at local community colleges and universities. According to the IT professionals, there are many excellent tutorials offered online – for both the beginner and the well versed Geek – but the website methods for learning how to build text in iOS does not allow for needed feedback. In turn, the online instructors simply say that practice makes perfect, and do not get discouraged because creating properties is still somewhat complex for the layman.

Understanding Words on a Home Screen

Another aspect of building text in iOS is linked to the use of techno-speak that IT technicians often joke about in the classroom as "giving them power over the masses" in terms of having a language that only Steve Jobs would appreciate. Still, there are online users who shun traditional college computer training because they understand such tasks as connected objects for instance variables, or how to allow objective-C implementation code that targets a temp text box for useful tie ins to a storyboard object.

Thus, the so-called learning curve for building text in iOS should not be left to the novice or risk headaches, jokes an IT professional.

Mobile Operating System Explained

The goal of understanding how to build text in iOS is linked to how this mobile operating system first developed when users discovered the wonders of iOS screens on the iPhones and other smart screens. Because the iOS user interface is all about direct manipulation - using the simple act of touching a screen – the data signals and text need skills that are well beyond a typical writer. In fact, many IT experts like to joke online about finding an iOS text personal who can navigate all the legal ins and outs of building text that is not already protected by copy write law of one form or another.

Moreover, anyone who decides to produce text in iOS must first understand that the development of the applications or apps must adhere to either the 32-bit or 64-bit ARM architecture. Also, the modern view from IT professionals is most users do not enjoy using text-heavy applications because they want the app to do the work for them. Still, the reliance on the apps to do more human brain work is vexing to those who enjoy building text in iOS because there is a very limited space on a Smartphone screen for lots of words.

Thus, what is trending today – when it comes to building text in iOS – is non-words and codes being used to explain simple operations that were once simple terms that have become "too wordy," say iOS text designers.

Text Input Protocol Explained for iOS

While it took Apple's brainpower to design and build the text layout methods and architecture for iOS 6 and 7, the funny thing about this development is the focus on what fonts to use over how an app really interfaces with a user. For example, an IT expert – who teaches at a community college and often, provides tutorials online – said the difficult part of building text in iOS programming is not about attractive layouts, reader-friendly text or messing with the grid for design purposes, but deciding on Helvetica, Century or Times New Roman fonts.

It is the simple and small details of building text in iOS programming that often vexes designers, add IT experts.

Also, there are new trending methods for hiding behind the labels of iOS text that include:

- New methods for text folding
- New interactive text color
- The emergence of custom truncation

In general, it is now possible to create text in iOS to have the color of the text color change as a means of communicating to the user that the app is ready for another command or operation. This revolution in how text is being built means words will share power with new options that are viewed as dynamic elements but are in fact more text layout architecture methods that are best used by true experts in computer design. The bottom line is the developers of iOS text are worrying more about looks than content.

Brave New World of Building Text for iOS

When iOS text builders mention dynamic type and optical scaling they are more concerned about the possibilities of making text fonts bigger or smaller at the touch of an app, than using fonts to enhance the message of words on an app. While the goal is to use heavier font weights for type that needs to be very small, the savvy

IT professional will tell iOS text builder hopefuls to dynamic text does not seem to work with the usual custom embedded fonts that some designers seem to like for looks alone.

The solution is to always try and maximum the size of the fonts used in building text in iOS because the real goal is to always improve legibility for apps. The rationale is the top million seller apps on the market today all feature readable type that may change in size – depending on the screen being used – but the text is always legible and clear in its wording. The goal of any iOS text developer is also linked to making the application as attractive as possible. While it would seem that art, images and the use of color alone does the trick, the experts who have researched successful apps say it is apps with a nice Helvetica font and readable text that sell best.

Overall, there are many avenues and methods when it comes to the high-text task of creating text for iOS, but at the end of the day the developers of the method say it still comes down to the right words that are clear and dynamic in communicating a message or instruction for users.

ABOUT THE AUTHOR

Jason Scotts loves technology so when iOS programming was first launched he was eager to learn all that he could about it. As he started to learn more about the inner workings of the program, he started to see all the benefits that came with it and even some of the few shortcomings. The pros far outweighed the cons so he made the decision to spread the word on this technology.

There are many programs that can be used to do quite a number of things but to date, as Jason has discovered. iOS programming is unique and stands out from the rest. He explains why in his text.

www.ingramcontent.com/pod-product-compliance
Lightning Source LLC
Chambersburg PA
CBHW070905070326
40690CB00009B/2005